ALL ABOUT

The Human
BODY

ALL ABOUT
The Human
BODY

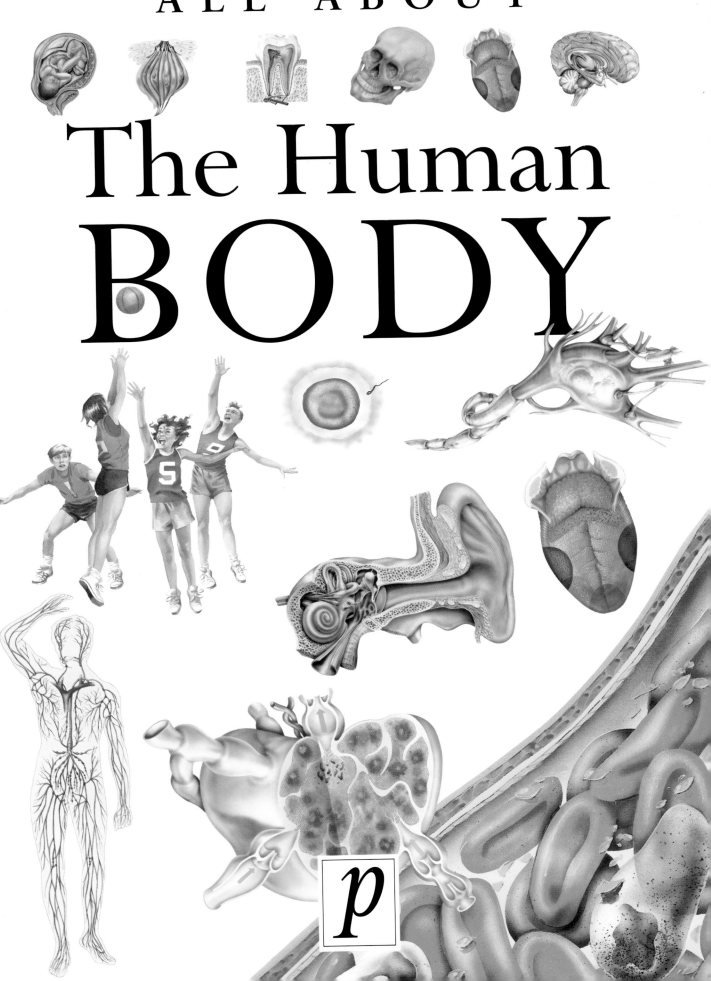

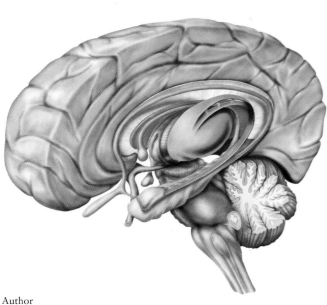

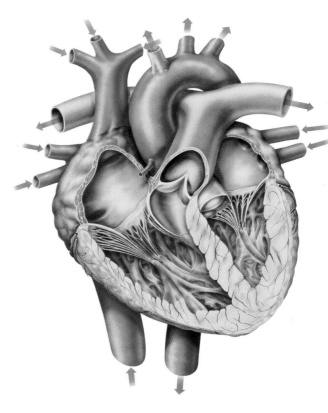

Author
Steve Parker

Designers
Diane Clouting and Phil Kay

Editor
Linda Sonntag

Project Management
Raje Airey and Liz Dalby

Artwork Commissioning
Susanne Grant

Picture Research
Janice Bracken and Kate Miles

Additional editorial help from
Lesley Cartlidge, Jenni Cozens, Libbe Mella and Ian Paulyn

Editorial Director
Paula Borton

Art Director
Clare Sleven

Director
Jim Miles

This is a Parragon Publishing Book
This edition published in 2000

Parragon Publishing, Queen Street House, 4 Queen Street, Bath, BA1 1HE, UK
Copyright © Parragon 1999

Produced by Miles Kelly Publishing Ltd
Bardfield Centre, Great Bardfield, Essex, England CM7 4SL

ISBN 0-75254-527-2

Printed in Singapore

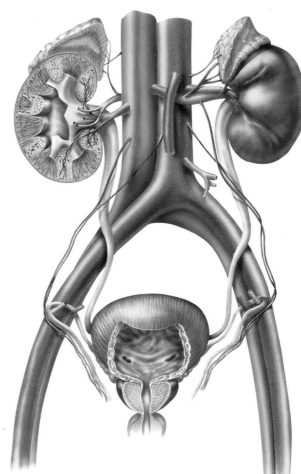

CONTENTS

HUMAN BODY

ALL ABOUT THE HUMAN BODY is divided into fifteen different topics, each covered by a double-page spread. On every spread, you can find some or all of the following:

- Main text to introduce the topic
- The main illustration, designed to inform about an important aspect of the topic
- Smaller illustrations with captions, to describe aspects of the topic in detail
- Photographs of unusual or specialized subjects
- Fact boxes and charts, containing interesting nuggets of information
- Biography boxes, about the scientists who have helped us to understand the way the human body works
- Projects and activities

BODY SYSTEMS

THE MOST STUDIED OBJECT in the whole of history is—the human body. More than 2,000 years ago, people began to wonder what was under its skin, and how all the parts worked. Doctors tried to find out why it fell ill, and why it sometimes got better. Artists, painters, and sculptors studied the body's shape to produce great works of art. People also asked endless questions about the body. How does it begin? How does it grow from a small and helpless baby into a large, strong, active adult? Why does it become old? Today, many of these questions have been answered, but the body remains a source of fascination and wonder. The body is where we live, move, work, play, eat, drink, think, and sleep. The following pages show the main parts of the body and how they work together to keep it alive for 70 years or more.

How the body is made up

The closer you look at the body, the more you find. The whole body seems like a single object or item. Yet it contains hundreds of organs. Each organ has millions of cells. On an even smaller scale, each cell has billions of molecules, and so on ....

Body cells

Body tissues

Body organs

Body systems

Whole body

Studying the body through history

The scientific study of the body's structure is called anatomy. The study of how it works is physiology. The first great anatomist was Claudius Galen of Ancient Rome. He looked after gladiators who fought terrible battles in the Coliseum, so he saw plenty of human insides. Today we use powerful microscopes and scanners to study the body. Yet there is still a great deal left to discover.

Skulls were worshipped in ancient times.

The human animal

There are almost six billion human bodies in the world. They all belong to one group, or species, *Homo sapiens*—"wise human." Inside, our bodies are very similar to those of many other animals—especially the great apes, the chimps and gorillas. Scientists agree that humans evolved from a prehistoric apelike ancestor, by changing gradually over millions of years.

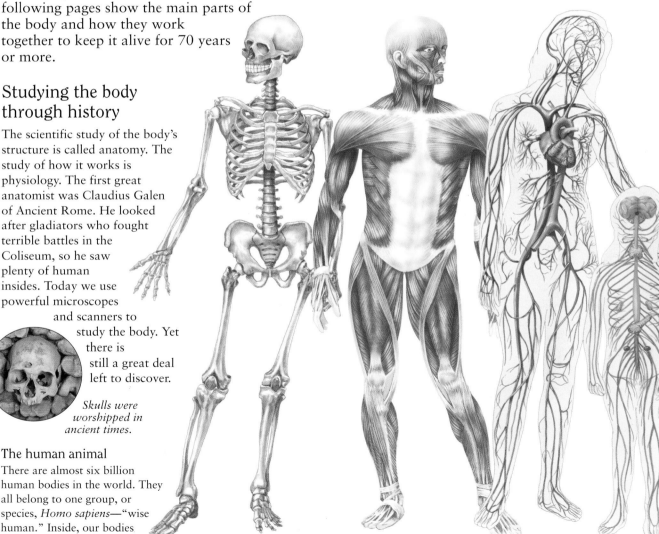

The skeletal system has 206 bones, strong yet light, and linked at joints. They form the body's flexible inner frame.

The muscular system has more than 600 individual muscles. Most pull on bones to cause bodily movements.

The circulatory system is the body's internal transport network, carrying vital substances such as nutrients.

The nervous and hormonal systems are in control of the body. They monitor and coordinate all parts and processes

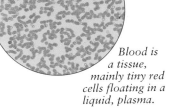

Blood is a tissue, mainly tiny red cells floating in a liquid, plasma.

Body systems

The body is made up of about [1]2 major systems. Each of these [do]es one very important job, [su]ch as supporting the body, [m]oving it or controlling it. [S]ome systems, such as the [n]ervous and lymphatic systems, [s]pread throughout the body. [O]thers are mainly in one place, [li]ke the respiratory system in [th]e head and chest.

Body organs

Each body system consists of several main body parts, called organs. For example, the digestive system is made up of the mouth, gullet, stomach, and intestines, plus the liver and pancreas. The circulatory system consists of the heart, the blood, and the immensely complex, branching network of tubes known as blood vessels.

Body tissues

An organ is made of substances called tissues. For instance, the heart has thick walls made of muscle tissue, a tough outer covering, and a thin and flexible lining of blood-proof tissue. It also has nerve tissue to coordinate its pumping movements, and connective tissue to hold all these other tissues together.

Body cells

Cells are the microscopic "building blocks" of the human body. Most kinds are about one-thirtieth of a millimeter across. In epithelial or covering tissue, the cells are box-shaped and joined together strongly, like bricks in a wall. In blood, which is also a kind of tissue, the cells float freely in a fluid. The body consists of more than 50 billion billion cells, of more than 200 different kinds.

Molecules

Each cell is made up of yet smaller parts, molecules. These form its outer skin or membrane, its control center or nucleus, and other parts. A red blood cell contains molecules of hemoglobin—the substance which carries oxygen—270 million of them.

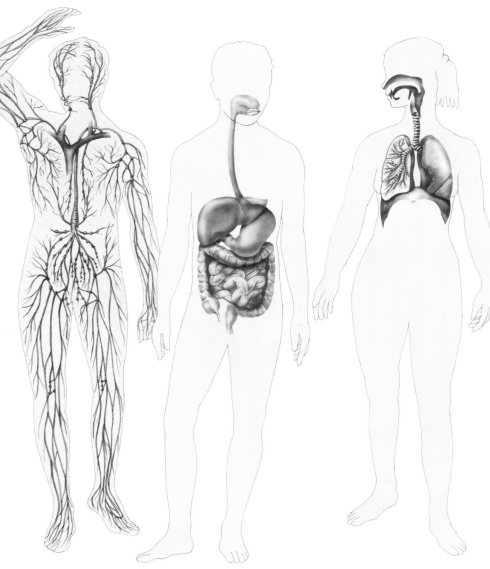

The lymphatic and immune systems are important in fighting disease and germs, and helping the body to recover from injury.

The digestive system takes in and breaks down food, to provide the raw materials for the body's growth and the energy to power its life processes.

The respiratory system breathes in air and takes from it oxygen (which all living things need to survive). It also gets rid of a waste, carbon dioxide.

The same but different

On the outside, people look very different. Some are short, others tall. Some have long legs, others have a long body. Some are light-skinned, others dark-skinned. We can change our appearance with the way we dress. But inside, every body has the same parts, such as a heart, lungs, muscle, and bones. Under the skin, we are all very much alike.

SKIN, HAIR, AND NAILS

WHEN WE LOOK AT a human body, nearly everything we see on the outside is dead. In fact, it ceased living weeks ago. This dead outer layer is skin, tough yet flexible. It protects the body from knocks, wear and tear, dirt and germs, rain and snow, and possibly harmful rays from the Sun. It also keeps in vital body fluids, salts and minerals. Just beneath this dead surface, skin is very much alive. It is one of the body's busiest organs. It contains tiny blood vessels to keep it nourished, and nerves that detect touch and pain. Skin also helps to control body temperature. When the body is too hot, skin goes flushed and sweaty, to help the body cool down. And the inner layers of skin continually replace the worn-away outer layers, so that the body stays covered and protected.

The size of skin

The skin of an adult person covers and area roughly the size of a single bed. It weighs around nine pounds. This makes it the body's largest single part, or organ. Skin varies in thickness from almost paper thin on the eyelids, to much thicker on the soles of the feet. Skin also responds to changing conditions. Where it is rubbed or pressed regularly, it becomes thicker, to give extra protection. Very thickened patches of skin are called calluses.

The upper layer of skin

The skin's upper layer is called the epidermis. Like the rest of the body, it is made of microscopic cells. At the base of the epidermis, these are very much alive. They multiply to make millions more cells every minute, pushing the cells above them toward the surface. As these cells move upward, they gradually fill with a hard substance called keratin, then die. About one month later they reach the surface, and are worn away.

The lower layer of skin

The skin's lower layer is the dermis. It contains millions of microscopic touch sensors that detect touch, pressure, heat, cold, and pain. Each has a nerve fiber linking it to the brain. The dermis also has tiny blood vessels, and microfibers of the tough substance collagen and the stretchy substance elastin. These make skin flexible yet strong.

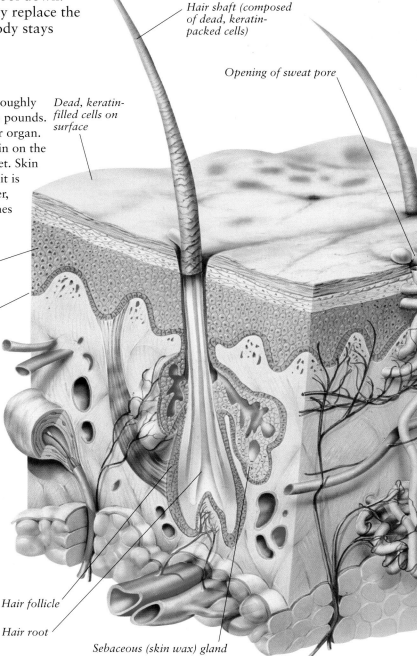

Hair shaft (composed of dead, keratin-packed cells)

Opening of sweat pore

Dead, keratin-filled cells on surface

Cells move upward, fill with keratin, and die

Rapidly multiplying cells at base of epidermis

Hair follicle

Hair root

Sebaceous (skin wax) gland

...ung and old skin

...baby's skin has had little wear and tear. ...s soft and smooth, and its hairs are tiny ...d bendy. Gradually the skin becomes ...ugher, and body hairs grow larger and ...cker. After many years, skin begins to ...e its natural stretchiness. Patches which ...nd a lot develop creases and wrinkles, ...e to a lack of elastin fibres.

Cortex

Medulla

Hollow
center

Cuticle

Hair and nails

An average person has about 100,000 head hairs. Like skin, they are made of dead cells. Each hair grows from a pit-shaped follicle in the skin. At the hair's base, or root, living cells multiply rapidly. Gradually they get pushed upward, cemented into a rod-like shape, filled with keratin, and die. Nails form in the same way, but as a flat plate. This grows from the nail root and slides slowly along, over the living skin below it.

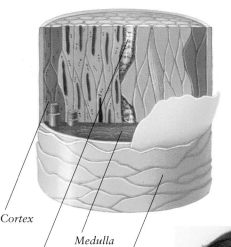

Curly or straight?

Hairs that are circular in cross-section tend to grow straight. If the hair is more oval, it grows naturally wavy along its length. If it has a C-shaped cross-section it is curly.

Sweat glands and pores

Sweat is a watery fluid made by around five million sweat glands scattered throughout the skin. Each gland is a tiny knot of tubes, connected to the surface by a corkscrew-like sweat duct. The duct opens at the skin's surface as a hole, the sweat pore (left). All the body's sweat ducts straightened out and joined together would stretch 6 miles. An average person produces about 3 fluid ounces of sweat in cool conditions. This rises thirteen times for an active person in hot conditions.

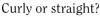

Layers of epidermis

Junction between epidermis and dermis

Sensory nerve endings for light touch

Sensory nerve endings for heavy pressure

Capillary

Sweat duct

Nerve fiber

Collagen and elastin fibers in dermis

Sensory nerve endings for changes in temperature

Sweat gland

Subcutaneous ("under-skin") fat layer

HOT OR COLD?

Skin can sense changes in temperature better than actual temperature. Fill three bowls with safely warm water, cool water, and cold water. Put one hand into the warm water and the other in the cool water. Then put both hands in the cold water. The hand that was in the warm water should feel colder than the hand in the cool water, because the temperature change was greater.

MUSCLES AND MOVEMENT

ALMOST HALF THE BODY'S weight is muscle. A muscle is a body part specialized to get shorter, or contract, when it receives nerve signals from the brain. Most muscles are long and strap-shaped. They taper at each end into ropelike tendons which attach firmly to bones. When the muscle contracts, it pulls the bone and moves that part of the body. This sounds simple, but the process of moving is incredibly complicated. There are more than 640 muscles, and they hardly ever work alone. They usually work in teams to pull, tilt, and twist several bones at once. Also, as one part moves, such as when you hold your arm out sideways, muscles in other parts need to work too. Your back and front muscles tense to take the strain, and your leg muscles shift weight to keep you balanced. The result is a smooth, coordinated movement, without your falling over!

Inside a muscle

A typical muscle is made of bundles of muscle fibers, or myofibers. Each fiber is slightly thinner than a human hair. In turn, each muscle fiber is a bundle of even thinner parts, muscle fibrils or myofibrils. And in turn again, each fibril contains bundles of long thin stringy substances known as actin and myosin. When a muscle contracts, the actins slide past the myosins, like rows of people pulling ropes. As millions of actins and myosins do this, the whole muscle gets shorter.

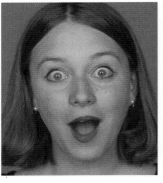 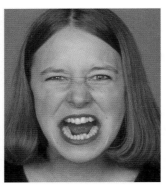

Making faces

About 60 muscles in the face, head, and neck produce our huge range of facial expressions. Some of these muscles are joined, not to bones, but to other muscles. For example, the frontalis muscles in your forehead can raise your eyebrows in a questioning way. Smiling is easier than frowning. A grin requires 20 muscles, while a grimace uses more than 40.

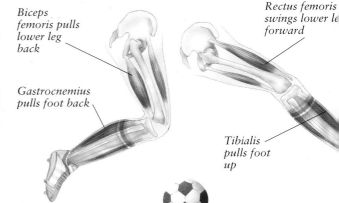

Biceps femoris pulls lower leg back

Gastrocnemius pulls foot back

Rectus femoris swings lower leg forward

Tibialis pulls foot up

Pull, not push

A muscle can get shorter and pull. But it cannot make a pushing force. So most muscles are arranged in opposing teams. One team pulls the body part one way. Then the other team pulls it back again. As each team pulls, the other relaxes and gets stretched. For example, muscles in the rear of the thigh pull the leg back at the hip and knee. Then opposing muscles in the front of the thigh quickly swing the leg forward and straighten the knee—KICK!

Power and coordination

In a complicated gymnastic movement, the brain must control almost every muscle in the body. This needs much practice, but gradually the brain part called the cerebellum takes over the control, and it becomes almost automatic.

Arms are ready to swing down on to floor just in front of feet

Legs swing up and over with lots of momentum

Legs continue to swing over on to floor and arms push up

Whole muscle

Fascicle (sheath of muscle fibres)

Single muscle fibre

Blood supply

Single muscle fibril

MUSCLES BIG AND SMALL

● The biggest muscle is the gluteus maximus, in the buttock. It pulls the leg backwards powerfully for walking and running.

● The longest muscle is the sartorius, from the outside of the hip, down and across to the inside of the knee. It twists and pulls the thigh outwards.

● The smallest muscle is the stapedius, deep in the ear. It is only 5 mm long and thinner than cotton thread. It is involved in hearing.

Sarcolemma covers muscle fibril

Muscle shapes

Many muscles, especially those in the arms and legs, are long and thin. As they contract, they bulge in the middle, at the part called the belly. But there are many other muscle shapes. The movements they produce depend on which other muscles are working at the same time, to tense or stabilize other parts of the body.

Banded pattern of muscle fibril

Bundles of actin and myosin

Actin molecule

Myosin molecule

Frontalis muscle raises eyebrow

Orbicularis oculis muscle is inside eyelids

Semispinalis muscle pulls back of head to look up

Deltoid muscle covers shoulder joint and helps to lift arm upwards

Biceps brachii muscle pulls forearm to bend elbow

Sartorius muscle

Gluteus maximus muscle

Body straightens and balances upright again

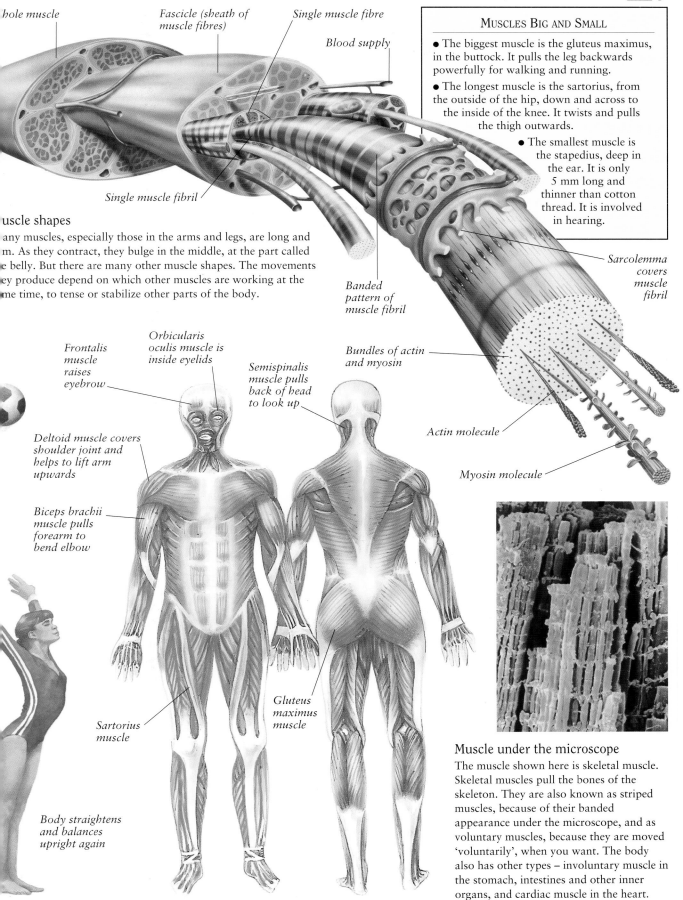

Muscle under the microscope

The muscle shown here is skeletal muscle. Skeletal muscles pull the bones of the skeleton. They are also known as striped muscles, because of their banded appearance under the microscope, and as voluntary muscles, because they are moved 'voluntarily', when you want. The body also has other types – involuntary muscle in the stomach, intestines and other inner organs, and cardiac muscle in the heart.

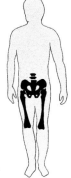

BONES AND JOINTS

MOST BODY PARTS, SUCH AS the guts, nerves and blood vessels, are soft and floppy. The body gets its strength and support from its skeleton. This is an inner framework of 206 bones. Many of them are linked at flexible joints, so they can move when they are pulled by muscles. There are many different shapes of bones, depending on the part of the body they support, and the jobs they do. Long bones in the arms and legs work like stiff girders, so we can reach and hold objects, and walk and run. The skull is a dome-shaped bone that surrounds and protects the delicate brain. The ribs form a moveable cage. This protects the heart and lungs within, yet also allows breathing movements. Without our bones, we would just flop on the floor like a heap of gelatin!

Flexible joints

Bones and joints, like other body parts, sta healthy if they are used properly. Exercise and activity can help to develop strong bones and smooth, flexible joints. But if a joint is regularly forced beyond its natural range of movement, it can eventually suffer problems of overuse such as pain and stiffness.

The skeleton

The skeleton has two main parts. The central towerlike axial skeleton consists of the skull, the backbone or vertebral column, and the ribs and breastbone in the chest. The appendicular skeleton is the limbs—the arms and legs. The bones in the legs are long and strong, to hold the body up and carry it when walking and running. The arm bones are very similar to the leg bones, but they are slimmer and lighter, and the joints are more flexible, for grasping and holding.

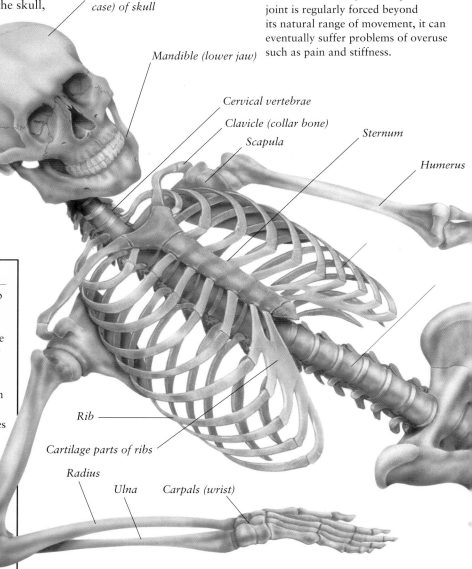

Cranium (brain case) of skull

Mandible (lower jaw)

Cervical vertebrae

Clavicle (collar bone)

Scapula

Sternum

Humerus

Rib

Cartilage parts of ribs

Radius

Ulna

Carpals (wrist)

KNOW YOUR BONES

● The largest bone is the pelvis, or hip bone. In fact it is made of six bones joined firmly together.

● The longest bone is the femur, in the thigh. It makes up almost one quarter of the body's total height.

● The smallest bone is the stirrup, deep in the ear. It is hardly larger than a grain of rice.

● The ears and nose do not have bones inside them. Their inner supports are cartilage, or gristle, which is lighter and bendier than bone. This is why the nose and ears are flexible.

● After death, cartilage rots faster than bone. This is why the skulls of skeletons have no nose or ears!

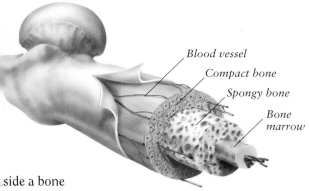

Blood vessel
Compact bone
Spongy bone
Bone marrow

Inside a bone

Old bones are dead, dry, and brittle. But in the body, bones are very much alive. They have their own nerves and blood vessels, and they do various jobs, such as storing body minerals. A typical bone has an outer layer of hard or compact bone, which is very strong, dense and tough. Inside this is a layer of spongy bone, which is like honeycomb, lighter and slightly flexible. In the middle of some bones is jelly-like bone marrow, where new cells are constantly being produced for the blood.

Joints

Inside a joint like the knee or elbow, the ends of the bones are covered with cartilage (gristle). This is shiny, smooth, and slippery, and allows the bones to slide past each other without wear or rubbing. The joint also contains a slippery liquid, synovial fluid, which works like oil to lubricate the movements.

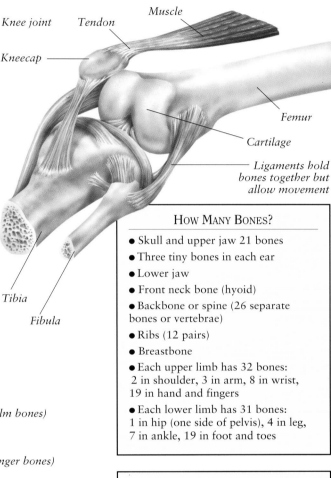

Knee joint
Kneecap
Tendon
Muscle
Femur
Cartilage
Ligaments hold bones together but allow movement
Tibia
Fibula

X-Rays

The invisible waves called X-rays can pass through the body's soft tissues, like muscles and nerves. They expose or darken a sheet of photographic film placed on the other side of the body. But the rays are stopped by dense tissues, mainly bone and cartilage. So these show up white. This X-ray reveals the bones in the wrist, palm, and fingers.

HOW MANY BONES?

- Skull and upper jaw 21 bones
- Three tiny bones in each ear
- Lower jaw
- Front neck bone (hyoid)
- Backbone or spine (26 separate bones or vertebrae)
- Ribs (12 pairs)
- Breastbone
- Each upper limb has 32 bones: 2 in shoulder, 3 in arm, 8 in wrist, 19 in hand and fingers
- Each lower limb has 31 bones: 1 in hip (one side of pelvis), 4 in leg, 7 in ankle, 19 in foot and toes

WHY ARE MANY BONES LIKE TUBES?

Roll and tape a sheet of cardboard into a tube. Fold another in half many times and tape it into a concertina. Fold and tape another sheet of card into a long box. Now try to bend each long shape. The tube should be strongest. This is why many bones are tubular.

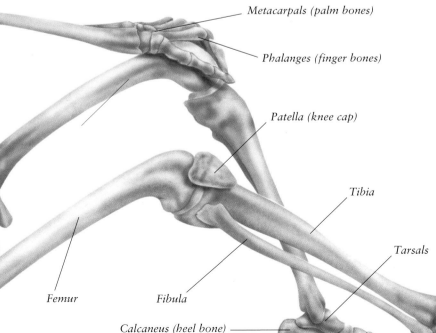

Metacarpals (palm bones)
Phalanges (finger bones)
Patella (knee cap)
Tibia
Tarsals
Femur
Fibula
Calcaneus (heel bone)
Phalanges (toe bones)

LUNGS AND BREATHING

EVERY ANIMAL BODY, including the human body, needs oxygen. This substance is a gas and makes up one-fifth of the air around us. Oxygen is needed because it is an essential part of the life processes inside the body, which break down the food we eat to release energy contained in it. The energy is used for growth, movement and every other body activity. The body system that obtains oxygen from air is called the respiratory system. Its main parts are the two lungs, inside the chest. The movements of breathing suck air into the lungs, where oxygen passes from the air into the blood. The blood flows away and carries the oxygen to all body parts. This is a continuous process, because the body cannot store oxygen for the future. So if breathing stops, the oxygen supply fails. After a few minutes, the body begins to die.

Nasal cavi

Pharynx (throa

Larynx (voicebo

Trachea (windpi

Cartilage rir

Le bronch

Left lu

Spa for hea

Ri

Bronchial tree of branching airways (bronchioles)

Right lung

Diaphragm

More and bigger breaths
During activity or exercise, muscles work harder. This means they use more energy. So they need more oxygen. The body responds by breathing faster and deeper, to take more air deeper into the lungs. At rest, the body takes in and breathes out about 10 litres of air each minute. After running a hard race, this amount can go up almost 10 times.

The respiratory system

This body system consists of the nose and throat, the windpipe or trachea in the neck, the main airways or bronch to each lung, the lungs themselves in the chest, and the main breathing muscle, the diaphragm. The lungs are light and spongy and contain smaller airways, bronchioles. The bronchioles branch and divide many times, becoming smalle and smaller, like an upside-down tree. Each breath draws in air through the nose, or the mouth, or both. Breathing through the nose helps to moisten and warm the air, while nose hairs filter out bits of dust. This helps to keep the lung clean and working efficiently.

Inside the lungs

Deep inside the lungs are more than 500 million microscopic air bubbles, which are called alveoli. These are surrounded by networks of microscopic blood vessels, the capillaries. Fresh air flows through the branching system of air tubes that fills the lungs, and into the alveoli. Oxygen from the air seeps easily through the ultra-thin walls of the alveoli and capillaries, into the blood, and is carried away round the body. Meanwhile one of the body's waste substances, the gas carbon dioxide, passes the opposite way, from the blood to the airways. The stale air is then breathed out.

Artery

Vein

Alveolus

Blood capillaries around alveolus

The mouth must be wide open when shouting, to allow the greatest air flow for maximum volume

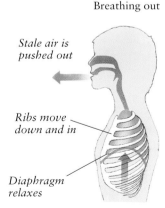

A supply of oxygen

All animals, including those in water, need oxygen. There is oxygen dissolved in water. Fish have gills, which are specialized to obtain the dissolved oxygen. Our lungs cannot obtain this dissolved oxygen. So divers take their own supply, in scuba tanks on their backs, or pumped down a long tube from the surface.

Noises from flowing air

Just below the throat, at the top of the windpipe, is the voicebox or larynx. Air flows through this during breathing. The larynx has two shelf-like folds of cartilage (gristle) in its walls, the vocal cords. In normal breathing these are wide apart. But they can be pulled almost together by voicebox muscles to form a narrow slit. As air flows through the slit it makes the cords shake fast or vibrate, and this produces a noise. Other muscles stretch the cords longer to make the noise higher-pitched, like a squeak. This is how we talk – and sing, shout and make many other noises, both deliberate and accidental!

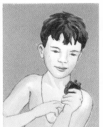
Breathing in

Fresh air is sucked in

Ribs move up and out

Diaphragm stretches

Breathing out

Stale air is pushed out

Ribs move down and in

Diaphragm relaxes

Breathing

Breathing involves two sets of muscles. One is a large sheet-like muscle in the base of the chest, the diaphragm. This pulls the lungs down. The other set is the intercostal muscles between the ribs, which pull them up and out. Together, these muscles stretch the lungs and make them suck in air. Then the muscles relax. The lungs spring back to their smaller size and push out the stale air.

FOOD AND DIGESTION

EVERY YEAR, AN AVERAGE PERSON eats about half a ton of food. But we do not put on half a ton of weight! We need to eat so much because there is a constant turnover within the body. Cells wear out and die, and new cells are made to replace them. The food we eat is broken down, or digested, into its many and various nutrients. Some of these are used as raw materials for growth, body maintenance and repair. Other nutrients are split apart to release the energy inside them, which is used for moving, breathing and hundreds of other life processes. Any undigested or leftover bits of food, along with various bodily wastes, are then removed from the body at the end of the digestive process and by excretion. Since the body is made from the food it eats, a balanced diet of good foods is important for health.

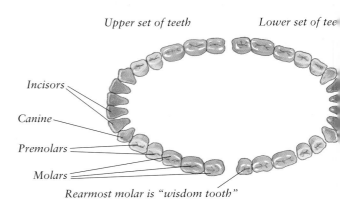

Upper set of teeth   *Lower set of tee*

Incisors

Canine

Premolars

Molars

Rearmost molar is "wisdom tooth"

Types of teeth

The adult human body has 32 teeth. The eight chisel-like incisors at the front bite food. The four pointed canines tear and rip it. Th eight premolars and twelve molars at the back of the mouth are flattened to crush and chew it. Before the adult teeth grow, the body has a first or "baby" set of 20 teeth. These begin to grow soon after birth, and are replaced by the adult teeth from the age of about six or seven years.

The digestive system

The main part of the digestive system is the digestive tract. This is like a long tube, some nine metres in total, through the middle of the body. It starts at the mouth, where food and drink enter, and finishes at the anus, where leftover food and wastes leave the body. Some parts of the tract are wide, such as the stomach. Others are narrow and coiled, like the small intestine. Each part has an important digestive task. Also involved in digestion are the liver and pancreas.

Fast food, slow eating

Thorough chewing mashes any kin of food into a pulp and makes it easier to digest in the stomach and intestines. Food that is bitten and swallowed with hardly any chewing cannot be digested so well.

Moving food

Lumps of food are pushed through the digestive tract by moving, wave-like contractions of the muscles in its wall, known as peristalsis.

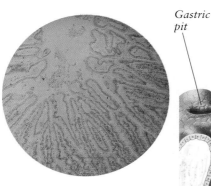

The villi of the small intestine under the microscope

Inside the intestine

The small intestine takes in or absorbs nutrients. The greater the area for absorption, the more effective this process. So the intestine's inner lining has thousands of tiny, finger-like villi. These increase its surface area almost 40 times compared to a flat lining.

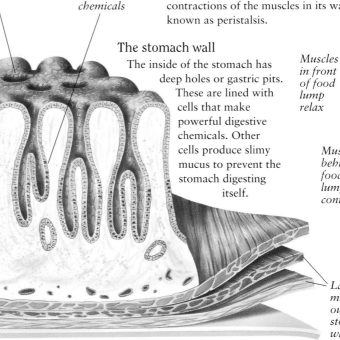

Gastric pit

Cells make digestive chemicals

The stomach wall

The inside of the stomach has deep holes or gastric pits. These are lined with cells that make powerful digestive chemicals. Other cells produce slimy mucus to prevent the stomach digesting itself.

Muscles in front of food lump relax

Muscles behind food lump contract

Layers of muscle in outer stomach wall

Layers o muscle i digestive tract wal

Mouth

Teeth bite off and chew food into a soft pulp that is easy to swallow. Chewing mixes the food with watery saliva, from six salivary glands around the mouth and face, to make it moist and slippery. The tongue moves the food around, positioning it between the teeth for thorough chewing and then separates off lumps and pushes them one by one into the throat for swallowing.

Liver

Blood from the intestines flows to the liver, carrying nutrients, vitamins, minerals, and other products from digestion. The liver is like a food-processing factory with more than 200 different jobs. It stores some nutrients, changes them from one form to another, and releases them into the blood, according to the activities and needs of the body.

Gall bladder

This small baglike part is tucked under the liver. It stores a fluid called bile, which is made in the liver. As food from a meal arrives in the small intestine, bile flows from the gall bladder along the bile duct into the intestine. It helps to digest fatty foods and also contains wastes for removal.

Small intestine

This part of the tract is narrow, but very long—about 20 feet. Here, more enzymes continue the chemical attack on the food. Finally the nutrients are small enough to pass through the lining of the small intestine, and into the blood. They are carried away to the liver and other body parts to be processed, stored and distributed.

Large intestine

Many useful substances in the leftovers, such as spare water and body minerals, are absorbed through the walls of the large intestine, back into the blood. The remains are formed into brown, semi-solid faeces, ready to be removed from the body.

Salivary glands

The three pairs of salivary glands, left and right, are under the tongue (sublingual), under the angle of the lower jaw (submandibular) and just in front of the ear. They produce about 3 pints of watery saliva (spit) each day. Some of this is stored, and released when chewing. The rest is released slowly and gradually, to keep the mouth, tongue and throat moist.

Gullet

The gullet, or esophagus, is a muscular tube. It takes food from the throat and pushes it down through the neck, and into the stomach. It moves food by waves of muscle contraction that pass along its length. These waves are called peristalsis, and they happen all the way along the digestive tract.

Stomach

The stomach has thick muscles in its wall. These contract to squash and mash the food into a sloppy soup. Also, the stomach lining produces strong digestive juices, called acids and enzymes. These attack the food in a chemical way, breaking down and dissolving its nutrients.

Pancreas

The pancreas is a wedge-shaped part behind the stomach. Like the stomach, it makes powerful digestive juices, called enzymes. When food enters the small intestine, the juices flow from the pancreas, along a thin tube, the pancreatic duct, into the small intestine. Along with enzymes made by the small intestine itself, the pancreatic enzymes help to digest food further.

Rectum and anus

The end of the large intestine and the next part of the tract, the rectum, store the faeces. These are finally squeezed through a ring of muscle, the anus, and out of the body.

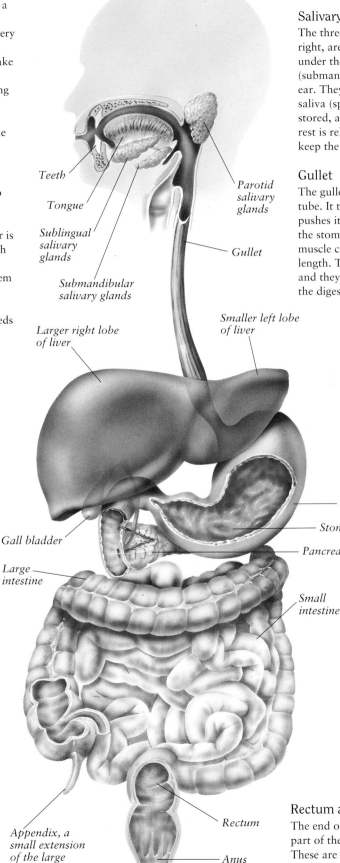

Teeth
Tongue
Sublingual salivary glands
Submandibular salivary glands
Parotid salivary glands
Gullet
Larger right lobe of liver
Smaller left lobe of liver
Stomach
Stomach lining
Pancreas
Gall bladder
Large intestine
Small intestine
Appendix, a small extension of the large intestine
Rectum
Anus

19

HEART, CIRCULATION AND BLOOD

THE HEART IS USED as a symbol of courage, strength and love. But it's not really involved in any of these. It is simply a hollow bag with muscular walls, which contract about once each second. The contractions squeeze blood from inside the heart, out through the bodywide network of tubes known as blood vessels. The heart, blood vessels and blood are known as the circulatory system, because the same blood goes round and round, or circulates. A typical person has about four to five litres of blood. This thick red fluid carries oxygen from the lungs, and nutrients from digestion, to every body part. It also collects body wastes and takes them to parts such as the kidneys and lungs for removal. In addition, blood carries substances called hormones, which control body processes, and antibodies to fight invading germs.

Inside the heart

Your heart is about the size of your clenched fist. It has thick muscular walls and is divided into two pumps by the septum. Each pump has two chambers. The upper, smaller, thin-walled atrium receives blood coming in from the veins. The blood flows through a one-way valve, which makes sure it always moves in the correct direction, into the larger, lower chamber. This is called the ventricle. It has thick, strong walls that contract to squeeze the blood through another valve, out into the arteries.

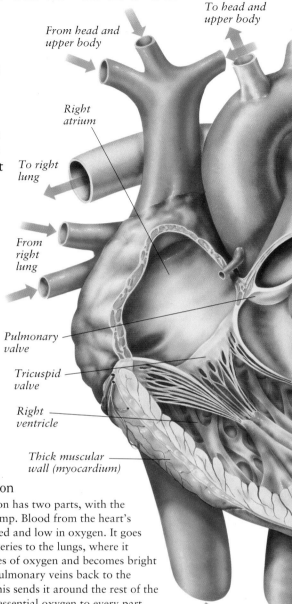

From head and upper body

To head and upper body

Right atrium

To right lung

From right lung

Pulmonary valve

Tricuspid valve

Right ventricle

Thick muscular wall (myocardium)

From lower body

To lower body

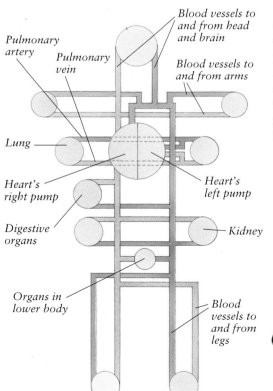

Pulmonary artery

Pulmonary vein

Blood vessels to and from head and brain

Blood vessels to and from arms

Lung

Heart's left pump

Heart's right pump

Digestive organs

Kidney

Organs in lower body

Blood vessels to and from legs

Two-part circulation

The body's circulation has two parts, with the heart as a double pump. Blood from the heart's right pump is dark red and low in oxygen. It goes along pulmonary arteries to the lungs, where it receives fresh supplies of oxygen and becomes bright red. It flows along pulmonary veins back to the heart's left pump. This sends it around the rest of the body, to deliver the essential oxygen to every part.

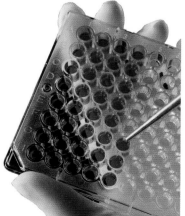

Blood transfusions

If someone loses a lot of blood, through a wound or other injury, one part of the treatment may be a blood transfusion. Blood previously given or donated by another person, the donor, is transferred or transfused into the ill person, or recipient. The blood must be matched. This means making sure it is the correct blood group for the recipient, by laboratory tests (left). Different people have different kinds, or groups, of blood. If blood of the wrong group is given, it may clump together and clot in the recipient's body, making the illness worse or even causing death.

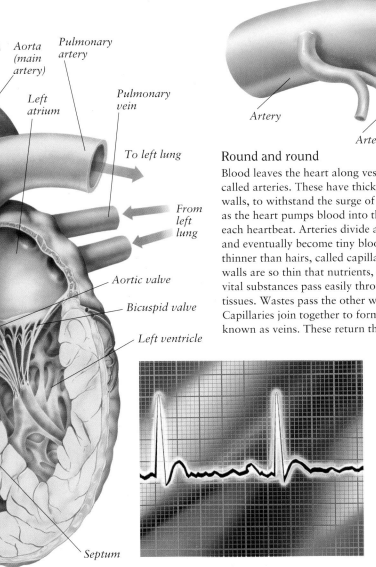

Aorta (main artery)

Pulmonary artery

Left atrium

Pulmonary vein

To left lung

From left lung

Aortic valve

Bicuspid valve

Left ventricle

Septum

Muscular wall

Endothelium (lining)

Lumen (space in middle)

Artery

Arterioles

Blood passes oxygen and nutrients to surrounding tissues

Capillary network

Venules

Vein

Wall of capillary

Round and round

Blood leaves the heart along vessels called arteries. These have thick, elastic walls, to withstand the surge of pressure as the heart pumps blood into them with each heartbeat. Arteries divide and branch and eventually become tiny blood vessels, thinner than hairs, called capillaries. Capillary walls are so thin that nutrients, oxygen and other vital substances pass easily through them, to surrounding tissues. Wastes pass the other way, into the blood. Capillaries join together to form larger, thin-walled vessels known as veins. These return the blood to the heart.

Heartbeats on screen

As the heart beats, it produces tiny natural pulses of electricity which ripple through body tissues. These pulses can be picked up by metal sensors, called electrodes, pressed on to the skin. A medical machine called the ECG, electro-cardiograph, can display the pulses as spiky lines on a monitor screen or strip of paper. The exact shape of the line tells doctors about the health of the heart.

Blood under the microscope

About half the volume of blood is a pale yellow, watery fluid called plasma. This contains dissolved nutrients, body minerals, salts, hormones and dozens of other substances. The rest of blood is microscopic cells. Donut-shaped red cells pick up vital oxygen in the lungs and carry it to the tissues. White cells are larger and help to protect the body against germs and illness. Platelets are tiny and help blood to clot, to seal a cut or wound. In one drop of blood there are about 5 million red cells, 300,000 platelets, and 10,000 white cells.

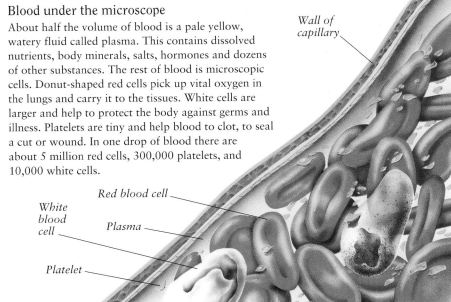

Red blood cell

White blood cell

Plasma

Platelet

SEE YOUR PULSE!

The heart must pump blood very forcefully, especially up to the head, to overcome the downward pull of gravity. This powerful pumping motion can be felt, using the fingertips as shown, just beneath the thin skin inside your wrist.

To see the effect of your pulse, attach a drinking straw to the inside of your wrist with plasticene. Hold your hand still and you can see that the straw moves in time with the rhythm of your pulse.

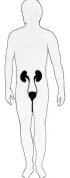

KIDNEYS AND EXCRETION

HUNDREDS OF LIFE PROCESSES inside the body produce many kinds of by-products and wastes. Removal of these is called excretion. It happens in several ways. The lungs breathe out carbon dioxide, a waste made when food nutrients are broken down to release their energy. The sweat that oozes on to the skin also contains some waste salts and minerals. But the chief method of removing wastes from inside the body is by the excretory system. Its main parts are the two kidneys. They "clean" blood passing through them, filtering out wastes and unwanted substances, to form the fluid called urine. This is stored in a stretchy bag, the bladder, until it's convenient to remove it from the body. Strangely, the removal of solid waste from the end of the digestive tract is not true excretion. These waste substances have been inside the tract, but not actually inside body tissues.

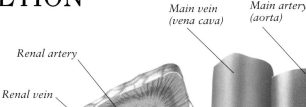

Main vein (vena cava)

Main artery (aorta)

Renal artery

Renal vein

Cortex

Medulla

Renal pelvis

The excretory or urinary system

The two kidneys are on either side of the backbone, shielded by the lowermost ribs. In most people, the left kidney is slightly higher than the right one. Each receives a huge supply of blood along its wide renal artery. This blood is filtered and cleaned and returns along the renal vein. Wastes and excess water collect as the pale yellow liquid, urine. This trickles from each kidney along a pipe called the ureter, down to the bladder in the lower body. When the bladder is full, the urine is released from it and passed along another tube, the urethra, out of the body.

Inside a kidney

A kidney has two main layers, an outer cortex and an inner medulla. The cortex contains about one million microscopic filters. These remove water and wastes from the blood. In the medulla, some of the water is taken back into the blood, according to the body's needs and water balance. Urine dribbles slowly but continuously into the kidney's central chamber, the renal pelvis. Then it passes along the urethra by the muscular squeezing process known as peristalsis, down the ureter to the bladder.

Bladder lining

Bladder covering (tunica serosa)

Main blood vessels down into right leg

Cystogram

Body fluids like blood and urine do not normally show up on an X-ray image. But in the X-ray called a cystogram, a person drinks a special harmless substance called contrast medium, which does show up. As it travels around in the blood and becomes concentrated in the urine, it reveals the shapes of the blood vessels and urine tubes.

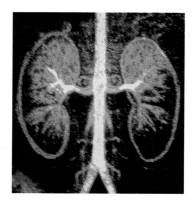

Muscles in bladder wall

Ureter inlet into bladder

Prostate gland (part of male organs)

Urethra

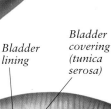

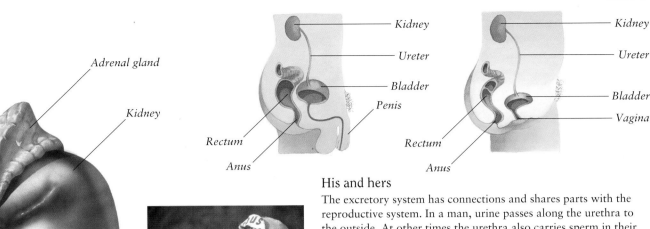

Adrenal gland

Kidney

Ureter

Kidney — Ureter — Bladder — Penis

Rectum

Anus

Kidney — Ureter — Bladder — Vagina

Rectum

Anus

His and hers

The excretory system has connections and shares parts with the reproductive system. In a man, urine passes along the urethra to the outside. At other times the urethra also carries sperm in their fluid, from the glands called the testes to the outside. The urethra runs along inside the penis, and the urine or the sperm in their fluid pass out of the end. So the reproductive and urinary openings from the body are the same. In a woman, the urethra opens to the outside just in front of the birth canal or reproductive opening, the vagina.

Kidneys in control

The kidneys not only remove wastes from blood. They also control or regulate the amount of water, minerals and salts in the blood—and so in the whole body. If you drink a lot, the water passes into the blood and makes it weaker, or more dilute. So the kidneys remove the extra water and get rid of it as extra urine, which is very pale and watery. If you do not drink much, or you sweat lots on a hot day, the kidneys conserve water in the body by removing less of it from the blood. So there is less urine.

One in a million micro-filters

The kidney's microscopic filters are called nephrons. Each has a knotty tangle of capillaries (tiny blood vessels) called the glomerulus (right), surrounded by a double-layered cup, the glomerular capsule. As blood flows through the glomerulus, wastes and water, plus some useful salts and minerals, are squeezed into the cup. They trickle from the cup along a tiny but long and winding tube, the nephron loop. This is also surrounded by capillaries. Useful substances and the required amount of water pass back into the blood. Wastes and excess water flow on, along larger tubes called collecting ducts, and form urine.

THE INS AND OUTS OF WATER

Ins On a typical day, an average person takes in about 80 fluid ounces of water. Around 50 fluid ounces is drinks, and 30 fluid ounces is in foods. Also, chemical processes inside body tissues actually make about 10 fluid ounces of water daily. This is called the water of metabolism. So, in total, the body's daily intake of water is around 90 fluid ounces.

Outs So that the body does not swell up like a wet balloon, it must get rid of as much water as it takes in. On the same typical day it loses 55 fluid ounces of water in urine, 7 fluid ounces in feces (digestive wastes), 18 fluid ounces as natural, slow, continual sweating, and 10 fluid ounces as water vapour in breathed-out air. So the body gives out as much water as it takes in: 90 fluid ounces.

In   Out

Metabolism

Food

Drinks

Water vapour

Sweat

Feces

Urine

HORMONAL SYSTEM

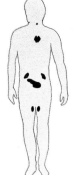

THE BODY HAS TWO overall control systems, to make sure that its many parts and processes work together smoothly. One is the nervous system. This is based on tiny pulses of electricity called nerve signals. The other is the hormonal or endocrine system. This is based on natural body substances called hormones. There are more than 50 different hormones, and they are all chemicals that 'carry messages'. They are made in hormonal or endocrine glands, and travel around the body in the blood system. Each type of hormone affects a certain part or parts of the body, like the heart, liver or stomach. These parts are called its target tissues or organs. The hormone makes its targets work faster (or slower). The higher the level of hormone in the blood, the faster (or slower) its target parts work. In this way hormones control or regulate the activities of many body parts.

Hormone-making glands

The 10 or so main hormone-making glands are found throughout the head and body, and are shown opposite. They release their hormones directly into the blood flowing through them, rather than along special tubes or ducts. So they are also called ductless or endocrine glands, as opposed to the ducted or exocrine glands like the tiny sweat glands in the skin. However, the pancreas in the upper left abdomen is both exocrine and endocrine. It makes digestive juices that flow along its duct into the intestines, and it produces hormones that pass directly into the blood flowing through this gland.

Scary ride

A rollercoaster ride brings mixed feelings of worry, fear, excitement, even great pleasure! The hormone system is working overtime as various endocrine glands release extra amounts of their hormones into the blood. Along with the actions of the nerve system, they make the heart pound, the muscles tense, the skin sweat and the skin turn pale. Some of these reactions are due to the hormone adrenaline (see opposite). Control by the nerves happens quickly, second by second. The effects of hormones last longer as these chemicals continue to circulate in the blood. This is why the excitement of the ride takes many minutes to die away as the heartbeat gradually slows down, muscles relax and skin returns to normal.

A rollercoaster ride gives you pale, sweaty skin, a pounding heart and tense muscles. These symptoms are caused by a rush of adrenaline, the body's reaction to a scary ride

Hypothalamus

Part of the brain just above the pituitary, it works with it to coordinate the hormone and nerve systems.

Pituitary gland

Makes about 10 hormones that control other hormonal glands; also regulates general body activities and growth.

Thyroid

Makes two hormones, thyroxine and tri-iodothyronine, that affect the speed of chemical processes inside cells, and another hormone, calcitonin, which regulates the mineral calcium.

Pancreas

Makes powerful enzyme-containing juices for the digestive system, d the hormones insulin and ucagon, which regulate ergy-rich sugar (glucose) in e blood.

Adrenal

The outer part (cortex) makes the body's natural steroid hormones, which affect levels of water, salts and minerals and help the body cope with stress and disease. The inner part (medulla) makes adrenaline (see above right).

Kidneys

Renin and other hormones help to balance the amounts of water, minerals and salts in blood d body fluids.

Stomach and intestines

Make hormones that work to coordinate the lengthy process of digestion.

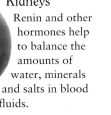

aries (female sex glands)

ake hormones that affect owth, including at the time of berty, when girl develops easts and gins the roductive menstrual cle of egg ripening.

Testes (male sex glands)

Make hormones that affect growth, including at the time of berty, when a boy develops cial hair and a deeper voice, d begins to produce sperm.

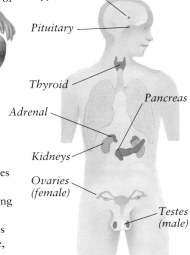

Hypothalamus

Pituitary

Thyroid

Adrenal

Pancreas

Kidneys

Ovaries (female)

Testes (male)

WHAT DOES ADRENALINE DO?

When the adrenal glands release adrenaline in a fight or flight situation, the following things happen to the body to enable it to react properly:

- Blood pressure rises
- Muscles receive extra blood
- Digestive and other organs receive less blood
- Skin receives less blood and so goes pale
- Heart rate increases
- Breathing rate increases
- Tiny airways in lungs (bronchioles) widen
- Liver releases high-energy sugar into bloodstream

Ready for action!

In a stressful or frightening situation, the brain tells the adrenal glands to release their hormone adrenaline. This acts in seconds to increase blood pressure, heart rate, the level of blood sugar, and the blood supply to muscles. These changes prepare the body for physical action, to cope with stress or danger. It's called the 'fight or flight' reaction because the body is prepared to fight or confront danger — or to flee (run away from) it.

TREATMENT FOR DIABETES

In the condition called diabetes, the pancreas does not make enough of the hormone insulin. This means body cells cannot take in blood sugar for energy. The sugar builds up in the blood and is removed by the kidneys. But it needs water to be dissolved in, so lots of water is also removed. The result is that someone with diabetes becomes thirsty and needs to drink large quantities of water, and also produces large amounts of urine containing lots of dissolved sugar.

In 1921 two research workers, Charles Best and Frederick Banting, carried out experiments in Ontario, Canada. They injected the hormone insulin, which had been purified from cattle pancreases, into a dog which had diabetes. The dog survived, the experiments were successful, and injections of insulin became a standard treatment for diabetes.

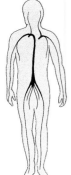

BODY DEFENSES

EVEN IN A CLEAN AND SPOTLESS place, there are probably a few germs. These microscopic living things, such as bacteria and viruses, get on to our skin, and into the food we eat, the drinks we consume, and the air we breathe. They may also get into the body through a cut or wound. If enough germs get into the body, they can start to multiply and cause problems. This is an infection. But the body has several sets of defenses against germs. These include the skin, the moist germ-trapping linings of the breathing and digestive passageways, the way blood clots to seal wounds and leaks, white cells and other substances in the blood, the thymus gland in the chest, and small lymph nodes or glands spread all over the body. Together, all of these parts form the body's immune defense system.

The body's germ killers

The body's immune system includes several kinds of white cells in blood, body fluids, and lymph nodes. Some are phagocytes, always on the lookout for germs such as viruse and bacteria. If a phagocyte finds a germ, it flows around it and "eats" it alive (see opposite). Other white cells, called lymphocytes, make body substances called antibodies. Thes stick on to the germs and make them helpless or burst open The white cells recognize germs and other invading substances because the invaders have foreign or "nonself" substances, antigens, on their surfaces. Antibodies work by attaching to antigens, like keys fitting into locks, to destroy the germ. Also, certain white cells "remember" the identity the germs. If the same type of germ tries to invade the body in the future, the immune system recognizes it at once and kills it, before it can multiply. This natural protection agains an infection second time around is called immunity.

White blood cells

The Human Immunodeficiency Virus, HIV, attacks the body's immune system—especially certain kinds of white blood cells. These are unable to make antibodies and fight back against the virus, in the normal way. The result is damage or deficiency to the immune system, causing the condition AIDS, Acquired Immuno-Deficiency Syndrome.

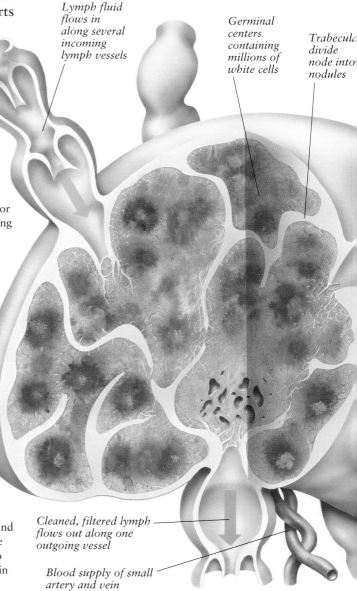

Lymph fluid flows in along several incoming lymph vessels

Germinal centers containing millions of white cells

Trabecula divide node into nodules

Cleaned, filtered lymph flows out along one outgoing vessel

Blood supply of small artery and vein

Cuts and clots

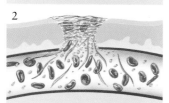

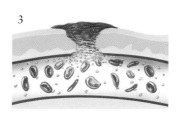

A wound or cut in the skin leaks blood from the damaged blood vessels (1). Chemicals released from damaged cells and platelets make dissolved substances in the blood turn into a meshwork of microfibers, of the substance fibrin. This network traps blood cells (2). Gradually the meshwork hardens into a clump or clot that seals the leak. The clot then hardens and dries further into a protective scab (3). White cells arrive to attack any germs, and the skin begins to regrow and heal.

The lymphatic system

The lymphatic system is of a network of tubes called lymph vessels, filled with a milky fluid, lymph. At various places around the body—especially the neck, armpits, lower body, and groin—these form lumpy-looking lymph nodes. Lymph fluid comes from general body liquids around and between cells, and blood leaked from blood vessels. These collect in lymph vessels and ooze along very slowly. Like blood, lymph fluid delivers nutrients to body tissues, and collects their wastes. It also contains many white cells, which defend the body against germs. The lymph vessels come together as large lymph ducts that carry lymph fluid back into the blood system near the heart.

Cervical lymph nodes in neck

Main lymphatic ducts

Axillary lymph nodes in armpit

Iliac lymph nodes in groin

EDWARD JENNER

Edward Jenner (1749–1823), an English country physician, noted that people who had cowpox, a mild disease caught from cattle, seemed resistant to the far more serious smallpox. In 1796 he gave a local boy cowpox. Six weeks later, he deliberately infected the boy's body with smallpox. The boy survived and Jenner's work laid the basis for immunization.

Tonsils and adenoids

The tonsils are patches of lymph tissue at the upper rear part of the throat. They help to destroy foreign substances that are breathed in or swallowed. The adenoids are similar patches at the rear of the nasal cavity inside the nose.

Thymus

The thymus gland in the front of the chest is large during childhood, but shrinks away during adulthood. It helps certain white cells of the immune system, especially lymphocytes, to develop and play their part in the body's defenses.

Spleen

The spleen is just behind the stomach on the left side. It makes and stores various kinds of white cells, especially the phagocytes that "eat" germs. It also makes and stores red cells for the blood, and generally cleans and filters blood.

Eating germs

Several types of white blood cells act as phagocytes. They flow around (1), engulf (2), neutralize (3), and digest (4) any strange or foreign objects, such as germs, before moving on to search for others (5).

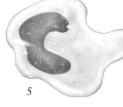
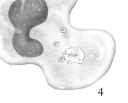

1
2
3
4
5

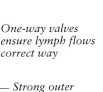

One-way valves ensure lymph flows correct way

Strong outer capsule

Inside a gland

When the body is ill with an infection, various glands swell up. Many of these are lymph nodes. In health they range from pea-sized to grape-sized, but in illness, they can be as big as golfballs. Lymph nodes contain billions of white cells, multiplying rapidly to fight the invading germs. During illness they fill with millions of extra white cells and also dead germs.

Protection for the future

Modern medicine can help the body to become immune or resistant to certain infections, without suffering them first time around. Specially weakened types of a germ, or some of its chemical products, are put into the body, usually by an injection or inoculation (vaccination). The germs are too weak to multiply and cause illness. But they do alert the immune system and produce resistance. This is known as immunization. It is usually carried out during childhood, to give lifelong protection. Which immunizations have you had?

EYES AND SEEING

PEOPLE OFTEN SAY: 'Seeing is believing.'
For most people, eyesight or vision is the
main sense. More than half of the
knowledge inside a brain enters through
the eyes – in the form of words,
diagrams, pictures and other visual
information. (Like now, as you read these
words.) So our eyes are our most
important sensory organs. A sense organ
is specialized to detect some aspect of the
surroundings – in the eye's case, light – and
produce tiny electrical nerve signals which travel
to the brain. The eye records what we see, but
does not interpret, analyse or understand it.
This happens in the sight centres at the
lower rear of the brain. Here, the nerve
signals from the eyes are decoded and
analysed. The result is that we see the
outside world in our 'mind's eye' – in full
colour and three dimensions, showing the
tiniest movements, intricate patterns and
amazing details.

Inside the eye

The eyeball's tough, white outer layer is the sclera.
Inside this is a soft, blood-rich, nourishing layer, the
choroid. Within this, around the sides and back of the
eye, is the retina. This layer detects patterns of light rays
and turns them into nerve signals, which go along the
optic nerve to the brain. The bulk of the eyeball is filled
with a clear jelly, vitreous humour. At the front of the eye is
the dome-shaped cornea, through which light rays enter. They
pass through a hole, the pupil, in a ring of muscle, the iris.
Then the rays shine through the bulging lens, which bends or
focuses them to form a sharp, clear picture on the retina.

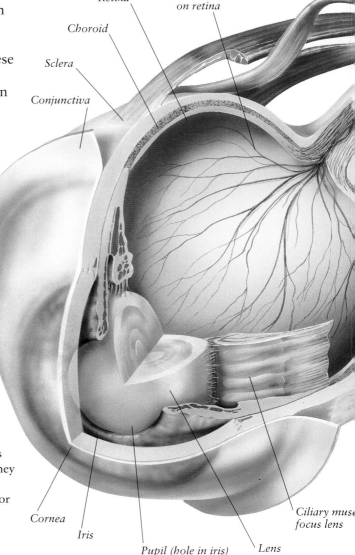

Retina
Blood vessels on retina
Choroid
Sclera
Conjunctiva
Cornea
Iris
Pupil (hole in iris)
Lens
Ciliary musc focus lens

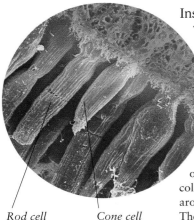

Rod cell   Cone cell

Inside the retina

The retina is about twice the
area of a thumbnail. Yet it
contains more than 130
million microscopic,
light-sensitive cells,
called rods and cones
because of their shape.
Rods number more than
120 million. They work
well in dim light, but detect
only shades of grey, with no
colours. Cones are concentrated
around the back of the retina.
They see colours and fine
details, but only work in
bright light.

Colour vision

There are three main kinds of cone cells in the eye. Each type is
most sensitive to a particular colour of
light – red, green or blue. The
thousands of delicate colours and
hues we see are combinations of
these three primary colours. In
some people, a particular type
of cone does not work
properly, or is absent. This
means colour vision cannot
work properly. The most
common problem is inability to
distinguish reds and greens, making
the star in the circle difficult to see.

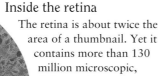

Careful with those eyes

The eyelids close when we blink to smear tear fluid over the eye, washing away dust and germs. The eyelashes also protect the eyes against floating bits and pieces. But eyes are at risk in very dusty or bright places, or when sawing, grinding, and working with chemicals or hot substances that may spray or splash. A mask, goggles or eye protectors are the sensible answer.

Eye-moving muscles

Optic nerve

Lachrymal or tear gland

Lachrymal ducts

Upper duct

Lower duct

Sclera (white of eye)

Iris

Pupil

Naso-lachrymal duct carries tear fluid into nose

Lachrymal sac

Look into my eyes …

The eyeball is partly hidden and protected in the bowl-shaped orbit, or eye socket, in the skull. From the front, we see only about one tenth of its surface. The white is the tough outer sclera. The colored disk is the iris. The hole in its center, the pupil, looks like a black dot. The iris, being muscle, can change shape. It makes the pupil wider in dim conditions, to take in as much light as possible. It contracts to make the pupil smaller in bright conditions, to prevent too-strong light entering the eye and damaging its delicate interior. The whole front of the eye is covered by a very thin, transparent membrane, the conjunctiva, which senses any dust or particles on the eye.

Area of full overlap

Left field of vision

Right field of vision

Area of partial overlap

How far away?

We judge distance or depth in many ways. We have two eyes, and so each sees an object from a slightly different viewpoint. The brain compares these two views. The more different they are, the closer the object. The brain also detects how far the eyes point inward. Again, the more they do this, the closer the object they are looking at. We also get clues from perspective—the way that lines, such as the two edges of a road, get closer together with distance. Another clue is parallax—as you move your head sideways, nearby objects pass in front of faraway ones. Also, colors fade with distance, and details become hazy. Relative size in another clue. A skyscraper that looks tiny is likely to be far away—not a tiny skyscraper that's very near!

CAN YOU BELIEVE YOUR EYES?

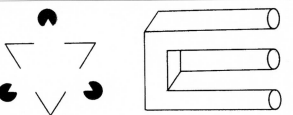

Sometimes our eyes see a scene, but the brain cannot make sense of it. This is usually because the scene could not exist in real life and three dimensions. It is drawn or painted on a flat surface, to trick the brain. For example, is there a clear triangle (left)? Can a C-shaped box have three holes (right)?

EARS, NOSE, AND TONGUE

THE HUMAN BODY HAS FIVE main senses. They are sight and touch (dealt with on previous pages), and the three shown here. Each detects a feature of the surroundings and produces nerve signals which travel along nerves to the brain. The ears detect sound waves traveling through the air. The nose detects floating smell or odor particles. The tongue detects flavor particles in foods and drinks. Like sight, these senses can give early warnings of danger, such as the sound of an approaching car, the smell of smoke from a fire, or the taste of rotting food. However, these five main senses are not the only ones. The body also has a range of inner sensors. Their job is to detect its temperature, the levels of chemicals and other substances in the blood, whether the stomach is full or not, and many other essential features.

Smell, or taste, or both?

When we eat a delicious meal, we experience many different sensations. But these are not all tastes and flavors. Many are smells. As we chew, food aromas waft from the back of the mouth, up into the nasal cavity, where they are detected as smells. If you eat food when your nose is blocked, by a cold or clothespin, food seems to have much less "taste." In fact, it tastes the same, but it lacks the accompanying smells.

Smell is as much a part of the pleasure of eating as taste.

On the tongue

The tongue is covered with dozens of pimple-like projections called papillae. These grip and move food when chewing. Around the sides of the papillae are about 10,000 microscopic taste buds. Different parts of the tongue are sensitive to different flavors: sweet, salt, sour, and bitter.

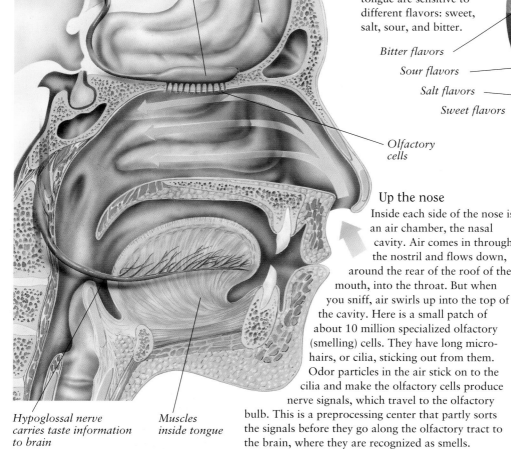

Olfactory tract   Olfactory bulb   Front of brain

Olfactory cells

Bitter flavors
Sour flavors
Salt flavors
Sweet flavors

Papil

Hypoglossal nerve carries taste information to brain

Muscles inside tongue

Up the nose

Inside each side of the nose is an air chamber, the nasal cavity. Air comes in through the nostril and flows down, around the rear of the roof of the mouth, into the throat. But when you sniff, air swirls up into the top of the cavity. Here is a small patch of about 10 million specialized olfactory (smelling) cells. They have long micro-hairs, or cilia, sticking out from them. Odor particles in the air stick on to the cilia and make the olfactory cells produce nerve signals, which travel to the olfactory bulb. This is a preprocessing center that partly sorts the signals before they go along the olfactory tract to the brain, where they are recognized as smells.

Taste bud

Each taste bud has some 30 gustatory (taste) cells clustered like the segments of an orange. The cells have micro-hairs, or cilia, sticking up on to the tongue's watery surface. As flavour particles touch the cili the gustatory cells produce nerve signals. These pass along nerves to the brain, where they are recognized as tastes.

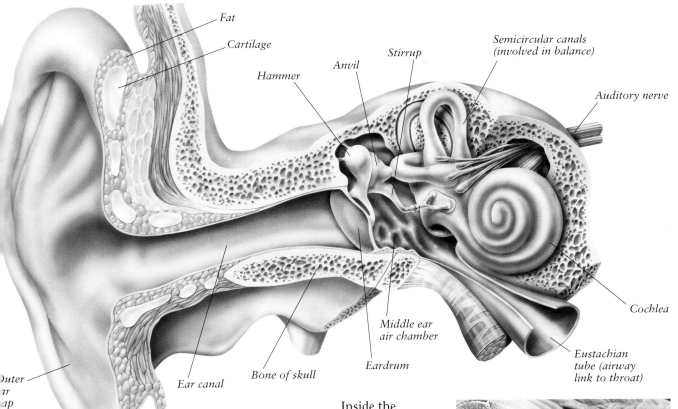

Fat

Cartilage

Hammer

Anvil

Stirrup

Semicircular canals (involved in balance)

Auditory nerve

Outer ear ap

Ear canal

Bone of skull

Eardrum

Middle ear air chamber

Eustachian tube (airway link to throat)

Cochlea

n the ear

ound waves funnel into the outer ear—the flap of skin and artilage on the side of the head. They pass along a narrow be, the ear canal, to a small patch of rubbery skin at its nd, the eardrum. The sound waves bounce off the eardrum nd make it shake to and fro, or vibrate. The eardrum is onnected to a row of three tiny bones linked together, the ammer, anvil and stirrup. The vibrations pass along these ones, like rattling the links of a chain. The stirrup presses gainst a small, fluid-filled, snail-shaped part, the cochlea, eep inside the ear. The vibrations pass as ripples into the uid inside the cochlea. Here, they shake thousands of tiny airs that stick into the fluid from hair cells. As the hairs hake, the hair cells make nerve signals, which go along the uditory nerve to the hearing centre of the brain.

Inside the cochlea

There are thousands of hair cells inside the cochlea. And each has dozens of micro-hairs sticking out from it (shown in yellow in the micro-photo, right). Some parts of the cochlea respond to high-pitched sounds, others to low-pitched ones.

Balance

Balance involves several sets of senses. Deep in the ear, chambers called the utricle and saccule detect the downward pull of gravity, so you know which way up your head is. Other inner ear organs called the semicircular canals contain fluid that swishes to and fro, telling you about head movements. Skin detects pressure as gravity presses body parts down. Eyes see horizontal and vertical lines, like floors and walls. The kinesthetic sense detects body position and posture (see left). All these sense inputs are analyzed in the brain, which sends out nerve signals to the muscles, to keep you balanced.

A SENSE OF POSITION

Yet another body sense is our inner sense of posture or body position. Try this. Stand up and close your eyes. Hold your arms out straight, to the front. How do you know when you have done this? You can "feel" the position of your arms and legs, body and head. Hundreds of micro-sensors in your muscles and joints detect the various amounts of stretch and strain in them. This inner sense of posture is called your proprioceptive or kinesthetic sense.

BRAIN AND NERVOUS SYSTEM

ALL YOUR THOUGHTS, feelings, memories, emotions, wishes, and dreams happen in one place inside your body—your brain. It's the place where you are conscious and aware of what's going on around you. It's the place where you think, imagine, have ideas, and daydream. Your brain is also the control center for your whole body. It tells your heart to beat, your lungs to breathe air, your stomach to digest food, and it controls hundreds of other inner body processes. And your brain makes your muscles work, so that you can move around and carry out skilled tasks such as drawing or riding a bicycle. The brain can do these amazing jobs because it is connected to all parts of your body, by an incredibly complicated network of nerves. These nerves look like thin pieces of shiny string. They carry messages, in the form of tiny electrical signals called nerve impulses, between the brain and every other body part.

Left cerebral hemisphere

Eye

The brain and nerve network

The brain fills the top half of the head. It is well protected by the strong skull bones around it. The brain's base tapers into a long, thick nerve called the spinal cord. Nerves branch from the cord and spread out through the body. These peripheral nerves divide and become smaller, and reach every part, even the fingertips and toes.

Parts of the brain

The brain looks like a giant, wrinkled walnut. Its most obvious part, making up about nine tenths of its whole size, is called the cerebrum. It is divided into two halves, known as cerebral hemispheres. Each hemisphere is covered by a pink-gray layer known as the cerebral cortex, which is where thinking happens. Under the cortex is a thick white layer, the cerebral medulla. At the rear of the brain is another wrinkled part, but much smaller than the cerebrum. This is known as the cerebellum. At the brain's base is a narrow part, the brain stem, which tapers into the spinal cord in the upper neck.

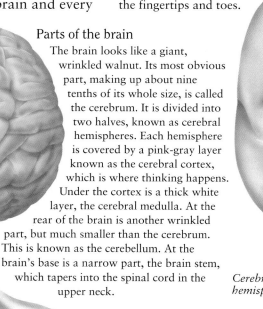

Cerebral hemisphere

Olfactory tract (to nose)

Insulation (myelin sheath) around nerve fiber

Main fiber (axon) of nerve cell

Body of nerve cell

Connection or synapse between nerve cells

Dendrites receive nerve signals from other nerve cells

Axon bulbs

How nerve cells communicate

Each microscopic nerve cell, or neuron, has a blob-shaped main part, the cell body, with thin, spiderlike dendrites and one much longer, wirelike nerve fiber or axon. The axon's branched ends have button-shaped axon bulbs which almost touch other nerve cells, at junctions known as synapses. Nerve signals travel along the axon and "jump" across the synapses to other nerve cells, at speeds of more than 300 feet per second.

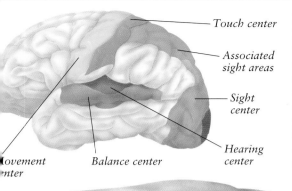

Touch center

Associated sight areas

Sight center

Hearing center

Movement center

Balance center

Inside the brain

Different patches or areas of the cortex receive nerve messages from the body's senses, such as the eyes and ears. These are cortical sensory centers. Another patch of the cortex, the motor area or center, sends out messages to the muscles so that the body can move.

Ultimate control

A person doing a skilled and complex task must concentrate hard. The brain makes thoughts and decisions every second, as information pours in from the eyes, ears and other senses, and signals are sent out to control hundreds of muscles with split-second timing.

Brain waves

Some electrical nerve signals in the brain pass though the skull bone to the skin. They can be detected by sensors and fed into an EEG (electro-encephalograph) machine, which displays them as wavy lines on a screen or paper strip. These "brain waves" can help doctors to discover any illness or problem affecting the brain.

Cerebellum controls muscle coordination and fine details of movement

Thalamus

Pituitary gland

Pons

Medulla oblongata

Spinal cord

Nerve cells

The brain is an immense web-like network of billions of microscopic nerve cells, called neurons. They pass tiny electrical signals amongst themselves. The total number of pathways that signals can take through the brain is unimaginably huge. The signals represent information coming from the senses along sensory nerves, thoughts and decisions and memories in the brain itself, and instructions going out to the muscles along motor nerves.

TEST YOUR MEMORY!

Memories are probably special pathways for nerve signals, within the massive nerve cell network of the brain. Try this memory test. Study the items below for 20 seconds. Close the book, write down all those you can remember, then open the book again to find your score. This tests short-term memory, which lasts for seconds or minutes. Next week, try to remember the items again, without looking! This tests long-term memory, which can last many years.

REPRODUCTION AND GROWTH

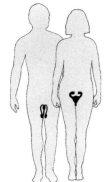

REPRODUCTION MEANS PRODUCING more of your own kind—that is, making babies. The parts of the body involved are called the reproductive system. They work much the same way in humans as in other mammals, such as chimps, cats, and horses. The female makes pinhead-sized egg cells called ova. The male makes microscopic, tadpole-shaped cells called sperm. During sex (sexual intercourse), sperm from the man pass into the reproductive system of the woman. One sperm may join with, or fertilize, an egg. If so, over nine months, the egg develops into a baby inside its mother's womb. It leaves the womb at birth. The baby continues to grow into a child, and then into a mature adult, when it can produce babies of its own.

Male reproductive system

The male sex glands are called testes. Each day they produc millions of tiny tadpole-shaped male sex cells, called sperm. These are stored in a tightly coiled tube, the epididymis. As new sperm are produced, the older sperm are gradually broken down and their nutrients recycled in the body. During sex, the man gently pushes his penis into the woman's vagina. After a while, sperm in their milky fluid ar forced along the vas deferens tubes, then along another tub the urethra, which is inside the penis. The sperm pass out o the male body and into the female reproductive system.

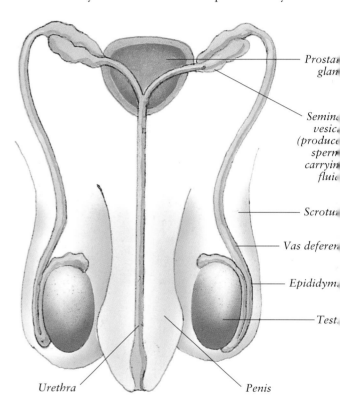

Prosta gland

Semin vesic (produc sperm carryin flui

Scrotu

Vas defere

Epididym

Test

Urethra

Penis

Egg meets sperm

A sperm carries genetic material (DNA) from the father. An egg is more than one hundred times larger than a sperm, but it contains an equal amount of genetic material from the mother. When egg and sperm join at fertilization, the two sets of genes come together, and a new individual is formed.

In the womb

A few hours after fertilization, the fertilized egg splits into two cells, then four, eight, and so on. They form a tiny ball which burrows into the nourishing womb lining. The cells continue to multiply, into hundreds, then thousands. They start to become different, as nerve cells, blood cells and many other types. Eight weeks after fertilization, a tiny baby has taken shape. It is only thumb-sized, but it has a brain, stomach, liver, eyes, and other main body parts, and its heart has started to beat.

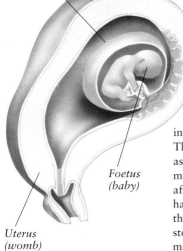

Amnion (protective membrane)

Foetus (baby)

Uterus (womb)

Ready for birth

Inside the womb, a baby is warm and protected. But it is surrounded by fluid, so it cannot breathe or eat. It obtains oxygen and nutrients from its mother through a plate-shaped part the placenta (afterbirth) in the womb lining. Blood flows from the baby's body along the curly umbilical cord to the placenta. Here it takes up oxygen and nutrients, gets rid of wastes, then flows back along the cord to the baby.

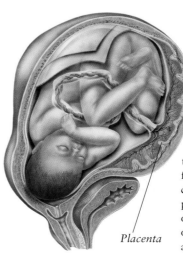

Placenta

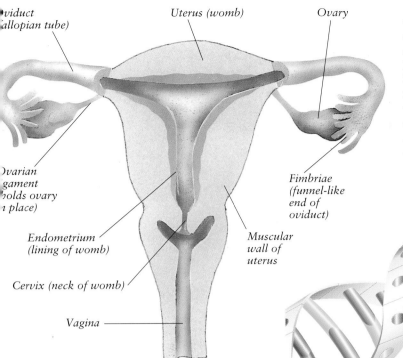

Oviduct (fallopian tube)

Uterus (womb)

Ovary

Ovarian ligament (holds ovary in place)

Fimbriae (funnel-like end of oviduct)

Endometrium (lining of womb)

Muscular wall of uterus

Cervix (neck of womb)

Vagina

Genes and DNA

Genes are instructions for the way the body grows and develops, and physical features such as colour of hair, skin and eyes. Genes are made of immensely long molecules of deoxyribonucleic acid, DNA. Each molecule has two long parts, backbones, twisted together in a long corkscrew shape called a double helix. The backbones are joined by four kinds of cross-links, known as nucleotides. The order of the cross-links is a chemical code for the genetic instructions.

Nucleotide cross-link

DNA backbone

Female reproductive system

The female sex glands are called ovaries. They contain tiny egg cells, or ova. Each month (usually) one egg cell becomes ripe and leaves the ovary. It passes along the fallopian tube and into the womb, or uterus. If sex takes place and the egg meets a sperm cell, it may be fertilized. The fertilized egg settles into the blood-rich lining of the womb and begins to develop into a baby. If there are no sperm, the egg and womb lining are lost through the cervix (the neck of the womb) along the passageway to the outside, the vagina. This bleeding is called menstruation, or a period. Once the period is over, the whole process of egg ripening and womb thickening begins again. This is called the menstrual cycle, and is controlled by hormones from the brain and ovaries.

Changing shape
As the body grows, it changes its proportions. A baby's head is much larger, relative to its body size, compared to an adult's head. This is shown clearly if bodies of different ages are drawn to the same size (below).

Inheritance
When an egg and sperm join together at fertilization, each contributes a full set of genes, which number between 100,000 and 200,000 per set. These genes, in the form of DNA, are in the nucleus (control centre) of the fertilized egg. They are copied when it divides, and at every division afterwards. So every body cell has two full sets or pairs of genes. But only one gene of the pair becomes active. This is why a child has some features passed on, or inherited, from the mother and others from the father.

The new arrival
Birth is an exhausting process for both mother and baby. The womb muscles contract powerfully to push the baby through the cervix, or neck of the womb, and along the vagina, or birth canal, into the outside world. Afterwards, mother and baby rest together and begin to get to know each other. The mother feeds the baby with natural milk from her breasts (mammary glands) or artificial milk from a bottle. This milk has all the energy and nutrients that the baby will need during its first few weeks of life.

The time after birth is very special, as mother and baby get used to each other by sight, sound, touch, and smell.

Growing up
The human body takes about 20 years to grow to its full physical size and maturity. The first stage is infancy, when the baby needs to be fed, cleaned, carried and cared for. Gradually it learns to smile, crawl, walk, and talk. During childhood we learn mental (mind-based) skills such as counting, reading and writing, and physical skills such as running, climbing, and riding a bicycle. During puberty the body grows very rapidly, and its sex organs become fully formed or mature.

THE HEALTHY HUMAN BODY

MOST PEOPLE ARE MOSTLY HEALTHY, for most of the time. We can help to improve our chances of a long and healthy life by understanding our bodies and how they work, and what makes them go wrong. Numerous medical studies and surveys have shown how our health is affected by various factors—what we eat and drink, how much and what type of physical exercise we take, the everyday activities we take part in, whether we take in harmful substances such as tobacco smoke, the way we relate to our family and friends, and even the way we organize our daily routine. Most of these factors are under our own control (more or less). So it's up to us to give our bodies the best chance. But there are also factors we cannot change so easily, such as where we live. And we cannot change some factors at all, like the genes we inherit from our parents, which may make us more likely to contract certain diseases.

Body weight

Being severely overweight, or obese, is associated with several kinds of illness. These include heart and blood-pressure problems, breathing difficulties, hormonal disorders such as diabetes, and joint aches and pains. Most people can control body weight by sensible healthy eating.

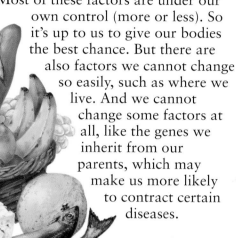

Food and drink

The body needs a varied diet. This includes proteins for growth, body maintenance and repair, starches (carbohydrates) for energy, and some fats for energy and maintenance, as well as vitamins and minerals for healthy teeth, bones, skin and other parts. Too much of any single food can cause problems. In general, plenty of fresh fruit and vegetables is helpful. Too much fat, especially animal fats from red meats and rich dairy produce, is not.

MAIN PARTS OF A HEALTHY DIET

Carbohydrates (starches and sugars) – In breads, pastas, potatoes, rice, and other grains, especially wholemeal or unrefined types.

Proteins – In plants such as peas, beans, and pulses, grains (and so in breads etc.), and in animal food such as meats, especially poultry and fish, and dairy produce such as eggs, milk, and cheese.

Fats – In plant oils such as olive, soy, sunflower, and safflower oils, also (but less desirable) in animal meats and dairy produce.

Fiber – In most fresh vegetables and fruits, also unrefined or wholemeal grains and their products such as wholemeal bread or rice.

Vitamins and minerals – In fresh fruit and vegetables mainly, also most other foods.

Before

Prior to strenuous exercise, it helps to do some stretching and warm-up exercises, to reduce the risk of sudden strains.

During

Correct clothing and equipment, such as well-fitting shoes, helps to avoid sprains and injuries.

Activity and exercise

The human body has evolved over thousands of years, to live a fairly active life—not to sit in front of a screen 18 hours a day. Most young people get exercise from sports and games. But many adults do not. After a medical fitness checkup, they may be helped by two or three periods of exercise weekly, each about 20-30 minutes long, and strenuous enough to make the lungs pant and the skin sweaty. But remember that some activities strain the same muscles continually, or stretch joints too much, and could lead to overuse injuries in the future.

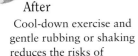

After
Cool-down exercise and gentle rubbing or shaking reduces the risks of stiffness and aches.

Rest and sleep

Lack of sleep can kill a person faster than lack of food. Trying to trick the body into needing less rest or sleep may result in headaches, lack of concentration, and being irritable and forgetful. This is part of the very important area of mental or mind health, which is put at risk by too much stress and worry.

Hygiene

Germs are almost everywhere. Regular hand-washing, especially after using the toilet, and before preparing foods or eating them, reduces the risk of infection—especially of taking germs into the digestive tract, when they could cause stomach or intestinal infection. Brushing teeth twice daily is an enormous help in preventing toothaches, tooth decay, and smelly breath. Dirty skin or clothes increase the risk of skin pimples, smells, and other skin problems.

Medical care

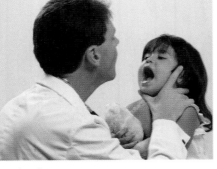

Most babies have regular health checks for the first months and years. These continue into childhood and adulthood in the form of visits to the optician for sight tests, to the dentist for tooth care, and to the doctor for checkups and screening tests. Immunizations in childhood also give protection against various infections. Any health problem is likely to be treated more effectively if it is reported to the doctor earlier, rather than later.

HARMFUL SUBSTANCES

Breathing in tobacco smoke is linked to many serious illnesses. The smoke contains vapors that condense or turn into thick tars. These clog the delicate lungs, increasing the risk of infections such as colds and bronchitis. A drug in tobacco smoke, called nicotine, is highly addictive. Cigarettes also contain cancer-causing substances and chemicals that affect blood pressure and damage blood vessels. There are many other drugs that also damage the health.

INDEX

ACKNOWLEDGMENTS

The publishers wish to thank the following artists who have contributed to this book:

Andrew Clark, Sally Launder, Mike Saunders, Guy Smith, Roger Stewart, Sue Stitt, Ross Watton.

The publishers wish to thank the following for supplying photographs for this book:

11 (C) Dr. Jeremy Burgess/Science Photo Library; 13 (BR) Quest/Science Photo Library; 21 (C) Mike Agliolo/Science Photo Library; 22 (BL) H. Sochurer/The Stock Market; 23 (CR) Prof. P. Motta/Dept. of Anatomy/University "La Sapienza", Rome/Science Photo Library; 25 (BR) British Diabetic Association; 28 (BL) Omikron/Science Photo Library; 31 (CR) Science Photo Library.

All other photographs taken from Miles Kelly archives.